CLOSE ENCOUNTERS

EXPLORING THE UNIVERSE
WITH THE HUBBLE SPACE TELESCOPE

ELAINE SCOTT

HYPERION BOOKS FOR CHILDREN
NEW YORK

Author's Note

NASA and ESA cooperated in building the Hubble Space Telescope. It is the size of a school bus and weighs about twelve tons. It orbits Earth at a speed of five miles per second. Inside are computers and instruments that serve as the electronic "eyes" of the astronomers here on earth. Two instruments are cameras—the Wide Field Planetary Camera II (WF/PC II), which takes pictures of objects at a relatively close range, and the Faint Object Camera (FOC). Instead of film, they use electronic detectors, similar to those used in home video cameras. Two other instruments aboard Hubble—the Faint Object Spectrograph and the Goddard High Resolution Spectrograph—spread starlight out into the colors of the spectrum. This information allows astronomers to study the star's composition, temperature, motion, and age. Computers on board the telescope convert all of the information Hubble gathers into long strings of numbers. These numbers are then beamed down to Earth as radio signals. At the Space Telescope Science Institute other computers turn the numbers back into pictures. The information transmitted in a single day is equal to the information contained in an entire encyclopedia!

Under the overall supervision of the Goddard Space Flight Center, observations made with the great observatory are done under the guidance of the Association of Universities for Research in Astronomy, or AURA, working with the Space Telescope Science Institute in Baltimore, Maryland. Although Hubble operates twenty-four hours a day, there are still not enough hours in the day to accommodate everyone who wants to use it. Astronomers must present proposals stating why they want to use the telescope to a selection board from the Space Telescope Science Institute. The proposals are submitted and evaluated once a year, and the competition for time on the telescope is brisk!

If you would like more information about the science being done by the Hubble Space Telescope, you may contact the Space Telescope Science Institute on-line at http://www.stsci.edu/top.html.

Acknowledgments

The following people made this book possible, and my deepest thanks go to them: Cheryl Gundy of the Space Telescope Science Institute, along with Dr. Laura Danly, Dr. Jeffrey Hayes, and Dr. Stephen Burrows, who took time from their busy observation schedules to talk with me. Astronaut/astronomers Dr. Jeffrey Hoffman and Captain Claude Nicollier not only repaired Hubble but also discussed its new observations with me. In addition, Jeff and Barbara Hoffman (a former librarian at the Kincaid School in Houston, Texas) gave the final manuscript expert readings—and corrections!

More thanks to the entire staff at the Jean V. Naggar Literary Agency and at Hyperion Books for Children, who have always supported my belief that stories rooted in science are as exciting as any rooted in the imagination.

Photo Credits

Page 3 courtesy of Betty Sullivan; page 5 (right) courtesy of Eric Schulzinger of Lockheed; pages 5 (left), 19, and 30 (inset) courtesy of the Space Telescope Science Institute; pages 39, 43, and 48/49, copyright © by Dana Berry.

All other photographs courtesy of NASA.

Printed in Singapore.
First Edition
1 3 5 7 9 10 8 6 4 2
This book is set in 13-point Meridien.
Designed by Mara Van Fleet.

Library of Congress Cataloging-in-Publication Data

Scott, Elaine, 1940–
Close encounters: exploring the universe with the Hubble Space Telescope / Elaine Scott.—1st ed.
p.   cm.
Includes Index
Summary: Describes what scientists have been able to deduce about the nature of our solar system and the universe based on data collected by the Hubble telescope.
ISBN 0-7868-0147-6 (trade)—ISBN 0-7868-2120-5 (lib. bdg.)
1. Outer space—Exploration—Juvenile literature. 2. Hubble Space Telescope—Juvenile literature.
[1. Outer space—Exploration. 2. Hubble Space Telescope. 3. Solar system.]   I. Title.
QB500.22.S36   1997
523—dc21   96-39104

CONTENTS

Edwin Hubble was born in Marshfield, Missouri. On the courthouse lawn, a replica of his namesake telescope is on permanent display.

INTRODUCTION

April 24, 1990. For decades, astronomers had waited for this day. The Hubble Space Telescope, a scientific instrument that some said was one of the greatest inventions of twentieth-century science, lay safely tucked inside the cargo bay of space shuttle *Discovery*, waiting to be launched into orbit 370 miles above the Earth. Once there, free from the distortions caused by Earth's atmosphere, the great observatory would peer into the farthest edges of the universe, seeking answers to questions scientists have asked for centuries. How old is the universe? How big is it? How are stars born? How do they die? Do black holes really exist? Have planets formed around other stars, the way our solar system formed around our Sun? And if they have, could life exist on one of those planets, too? The Hubble Space Telescope, or HST

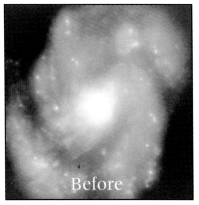

Before

for short, promised answers to these questions and many more. However, those who asked them in 1990 were bound for disappointment, for by June of that year—just two months after its launch—it became obvious that the Hubble was in trouble. The images it returned to Earth were fuzzy and out of focus.

There was talk that the telescope was ruined and could not be repaired; however, astronomers and engineers at the National Aeronautics and Space Administration (NASA) and the European Space Agency (ESA) rolled up their sleeves and went to work. Eventually a solution was found, and at last another launch day arrived. At 4:27 A.M. on December 2, 1993, the space shuttle *Endeavour* roared off its launchpad at Kennedy Space Center. Seven astronauts were on board. Their mission? Repair the Hubble. In a

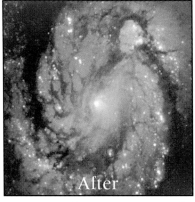

After

Following its repair, the telescope returned images of astonishing clarity.

series of complicated and risky space walks, the astronauts repaired existing instruments and installed new ones inside the telescope. Once the crew returned to Earth, the world held its breath as astronomers at the Space Telescope Science Institute in Baltimore, Maryland, fine-tuned the telescope by computer linkup. Just after midnight on December 18, 1993, the first images from the repaired Hubble began to return to Earth. Astronomers all over the  world were awestruck and filled with joy— the telescope had been fixed beyond their wildest expectations! Its vision was so clear it was like being in Washington, D.C., and seeing the light from a firefly in Tokyo, Japan—halfway around the world! And had there been two fireflies in Tokyo, it was as if you could see each one—even if they were only ten feet apart! Astronomer Dr. Laura Danly remembers the moment she first viewed the universe with Hubble's new "eyes." "I couldn't believe what I was seeing," she says today. "I was astonished. As I observed, I almost felt as if I were out there in space. It reminded me of the first time I put on glasses when I was a child, and suddenly, all the details of my world became clear. It's like that now with the Hubble. All of the glorious details of the universe are there."

 In the first few years since the repair, Hubble has looked within our solar system and beyond it. It has watched as stars are born and die, as a comet collided with Jupiter, and as matter spews around the edges of black holes. Although the telescope has answered many questions, it has raised others. Fortunately, more answers will come, as astronomers use the Hubble Space Telescope to do exciting science well into the next century.

Above left: Hubble Space Telescope astrophysicist Dr. Laura Danly poses with a model of the great observatory.

Above right: Dr. Christopher Burrows is credited with discovering the original flaw in the Hubble Space Telescope's mirror.

CHAPTER ONE

TELESCOPES THEN

Before there were telescopes—before there was even science—there were the stories ancient civilizations told in an attempt to explain the mysteries of nature. Egyptians once told the story of Ra, the sun god, who sailed his barge across the sky—which they thought of as a heavenly Nile River. As the story goes, Ra traveled the sky from east to west, bringing dawn, then day, and finally dusk. When Ra disappeared in the west and darkness fell, he was said to be traveling in the underworld, the land of the dead. And when an eclipse occurred, ancient Egyptians believed that a serpent had upset Ra's chariot!

Eventually human beings began to pay closer attention to the Sun and stars themselves, and to make predictions about their motions that they could test against what they could actually see. Science began to take the place of stories as a way to understand the world we live in. Astronomy, the branch of science that is based on observations of the stars, planets, asteroids, and other celestial bodies in the universe, is the world's first science. It was well established by 350 B.C., when Alexander the Great was conquering the known world. During that period, astronomers made startlingly accurate observations about the distance and size of the Sun and the Moon. However, as the science developed, mistakes were made, too. In A.D. 200, Ptolemy proposed that Earth was the center of the universe, and he developed a

AND NOW

Nicolaus Copernicus studied theology, medicine, mathematics, and astronomy. His argument that Earth is not the center of the universe changed the study of astronomy radically.

complicated theory to prove his point. For centuries, people accepted the Ptolemaic system of understanding how the universe was constructed, until Polish astronomer Nicolaus Copernicus (1473–1543) suggested that Earth and all the planets revolved around the Sun. At the time, few people believed Copernicus and his radical new idea.

Of course, all of these ancient astronomers had to rely upon their eyes to make their observations of the stars and planets. The telescope was not invented until the beginning of the seventeenth century. Credit for that invention usually goes to a Dutch optician named Hans Lippershey. As the story is told, one afternoon in 1608 Lippershey was toying with his lenses, holding one in front of another while he looked through both. As he peered through two lenses at once, the optician noticed that the objects he saw appeared larger and closer than they actually were. Since holding lenses apart with two hands was inconvenient, Lippershey placed them on opposite ends of a long tube, and the first telescope was created.

Jeffrey Hayes, an astrophysicist who works with the Hubble Space Telescope at the Space Telescope Science Institute, told more about Lippershey's invention. "The Dutch military heard about Lippershey's telescope and paid him 900 florins for it. They wanted to be able to identify foreign ships coming into the harbor before they got too close. They wanted to be able to tell if the ships were friends or foes." As he finished that anecdote, Jeff smiled and added, "The first telescopes weren't called spyglasses for nothing!"

News of Lippershey's telescope

spread quickly, and in Italy, a scientist named Galileo Galilei (1564–1642) soon made improvements to it. However, Galileo turned his telescope up instead of out, and used it to observe the heavens rather than spy on ships. Through his observations, Galileo discovered that the Moon had craters and mountains, that Jupiter had moons circling it, and that the Milky Way was not just a white smear in the night sky but was made up of millions of individual stars. Armed with this knowledge, Galileo endorsed Copernicus's theories that the planets orbit the Sun. Galileo's thinking was considered sacrilegious by those who said the Bible required people to believe that Earth was stationary and the center of the universe. The great astronomer was put before the Inquisition, a court that tried religious heretics, and was ordered into house arrest for proposing such controversial thoughts!

At about the same time that Galileo was using the telescope in Italy, the mathematician Johannes Kepler (1571–1630) was working with one in Germany. Kepler had been a student of the great Danish astronomer Tycho Brahe (1546–1601), and when Brahe died, Kepler continued to work with Brahe's observations. Kepler's use of a telescope enabled him to discover that planets did not orbit the Sun in circular patterns, as Copernicus and Galileo thought, but moved in oval, or elliptical, orbits instead. Kepler's three discoveries about the shape and speed of planetary orbits are called Kepler's laws, and they form part of the foundation of modern astronomy. The invention of the telescope revolutionized astronomy forever.

Most telescopes—including the Hubble Space Telescope—"see" by gathering light from the objects they study. Early telescopes collected light with a glass lens, which bends—or refracts—the light passing through it. Different colors of light, however, refract at different angles, and so the lenses in early telescopes spread white light out into the separate colors of the rainbow. Everything that Galileo and Kepler saw through their telescopes had an annoying colored edge to it. The brilliant English mathematician and astronomer Sir Isaac Newton (1643–1727) fixed that problem in 1668, when he invented a new and improved telescope.

Opposite page: Galileo was ordered into house arrest by the Inquisition after he wrote about his support of Copernicus.

Although Isaac Newton is often remembered for his work with gravity, he was also interested in optics—the scientific study of light. Thanks to his work with prisms, Newton understood what happened to light when it was refracted. He decided to use a mirror in his telescope to reflect the light rather than refract it. His idea worked, and the colored edges surrounding objects disappeared! Newton used his new, improved telescope to collect information, and then he developed a new branch of mathematics, called calculus, to analyze it. His theories led him to make the surprising suggestion that the same force that caused an apple to fall to the ground determined the path of planets around the Sun, an idea that eventually became known as Newton's theory of gravity or Newton's law of universal gravitation. Earlier, Kepler's laws had accurately described how the planets moved; Newton's law of universal gravitation explained why.

Newton's discoveries about the movements of planets and the force of gravity form the foundation for modern physics and the exciting explorations of the universe that are going on today. None of these explorations would have been possible without the use of telescopes. Today, there are several great telescopes located in observatories around the world. Most of them, including the Hubble telescope, use mirrors to gather light and are called reflecting telescopes.

Johannes Kepler was a professor of mathematics in Germany and later became imperial mathematician at Prague. After years of careful work, he developed Kepler's laws of planetary motion.

The Hubble Space Telescope was named in honor of Edwin Powell Hubble, an American astronomer who lived from 1889 to 1953. Until Hubble did his pioneering work with the telescopes at the Mount Wilson Observatory in Pasadena, California, most astronomers assumed our galaxy, the Milky Way, was the only one in the entire universe. Edwin Hubble's work led to the understanding we have today of the universe, a universe that contains hundreds of billions of galaxies, each one filled with hundreds of billions of stars. Our own space "neighborhood" contains at least twenty galaxies fairly close to ours. Astronomers call them the local group. The Andromeda Galaxy is the largest galaxy in the local group; the Milky Way comes in second.

In addition to proving that there was more than one galaxy in the universe, Dr. Hubble also discovered that almost everywhere we look, galaxies are moving away from each other. This discovery, now called Hubble's law, suggested that the universe and everything in it continues to expand outward from an initial explosion that is often called the big bang. Edwin Hubble also collected measurements that showed that the speed at which a galaxy races away into space is in direct proportion to its distance from us. In other words, the faster a galaxy is moving away from us, the farther away it is. And the farther away a galaxy is, the longer it will take its light to reach us here on Earth. After its repair, the Hubble Space Telescope captured light that began its journey toward Earth *billions* of years ago! It came from a star that had formed in the earliest days of the universe and was still moving out and away through space and time.

Scientists measure the incredible distances in the universe in terms of light-years. A light-year is the distance light travels in a year, about 5.9 trillion miles. In order to get a grasp of the kinds of distances astronomers deal with when they look out into space with Hubble, consider these facts: if you counted at the rate of one number per second, it would take eleven and a half days to count to a million, and thirty-two years to reach a billion, which is a thousand million. One trillion equals a thousand billion and looks like this: 1,000,000,000,000. Multiply 5.9 trillion times several billion and you can begin to grasp why the astronomers at the Space Telescope Science Institute were—and are— awestruck when they saw stars from this far away!

Edwin Hubble once said, "From our home on Earth, we look out into the distances and strive to imagine the sort of world into which we are born. . . . Our immediate neighbor-

hood we know rather intimately. But with increasing distance our knowledge fades rapidly." Now, thanks to the telescope named in his honor, astronomers are able to leave the neighborhood of our own solar system and go exploring billions of light-years away, toward the very edge of the universe. The telescope's journey of exploration is exciting and challenging. During its first few years in orbit, this is what Hubble saw.

Among his many other outstanding achievements, such as Hubble's law and the Hubble constant, astronomer Edwin Hubble was also responsible for classifying and naming galaxies as either spiral, elliptical, or irregular. Astronomers use those classifications today.

Inset: Galaxies cluster together, and so do stars. The smaller, white cluster of stars (bottom center) is younger than the yellow cluster. Although they look close together in this image, they are separated by two hundred light-years.

Hubble Space Telescope can take pictures of galaxies that formed less than one billion years after the big bang. Common spiral and elliptical galaxies, along with puzzling new kinds of galaxies, are shown in this image that covers a very small sample of the sky.

CHAPTER TWO

Most of us live in neighborhoods. A neighborhood is a small part of a larger city or town, which is, in turn, part of a state and then a country. Finally, our country is a part of our world—the planet Earth. But there are also neighborhoods that are literally out of this world, and we belong to them, as well. The nine planets and more than sixty moons, plus the millions of asteroids, comets, and micrometeoroids that orbit our star, the Sun, make up our solar system. The solar system is our immediate "neighborhood" within our celestial "city," the Milky Way. Although our solar system is 7.5 billion miles across, the Hubble Space Telescope can easily explore objects at its outer limits as well as those closer to Earth. One observation sent back new and intriguing information about our next-door (by astronomical standards) neighbor, Mars.

No other object in our solar system has captured our imagination more than this small red dot in the sky, so like our own planet Earth. In 1877, using a telescope that was modern for its day, Italian astronomer Giovanni Schiaparelli (1835–1910) observed something he called *canali*, the Italian word for "channels" or "grooves." Other scientists of the day speculated that these canals might have been built by intelligent beings. Scientists and nonscientists alike were excited.

THE NEIGHBORHOOD

In this image of Mars, its polar ice cap is clearly visible, though it is springtime on the planet. HST took the image on February 25, 1995, when the planet was approximately 65 million miles from Earth. The small red dot to the left in the image is an extinct volcano.

The Hubble Space Telescope is able to provide complete global coverage of Mars.

What if there was life on Mars? That question was debated for the next hundred years or so and was the basis for many science fiction books and films. The term Martian even found its way into our vocabulary and became a synonym for any extraterrestrial creature! Most of the speculation was finally put to rest in 1976, when the United States landed two unmanned Viking spacecrafts on Mars. Cameras inside the spacecrafts took snapshots of the planet, and instruments that were dropped to its surface recorded weather information to send back to Earth. The cameras revealed that what had appeared to be carefully constructed canals were simply dried-up riverbeds. And the weather instruments indicated the Martian climate was awful—extremely cold, with a sky full of pink iron oxide dust that gave the planet its rosy hue. Everyone agreed: Mars did not seem to be a hospitable place for any living creature—human or otherwise. However, almost twenty years later, pictures from the Hubble began to change scientific thinking—at least, about Mars's weather!

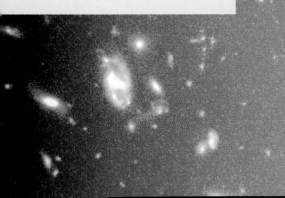

Valles Marineris Region
60° Longitude

Tharsis Region
160° Longitude

Syrtis Major Region
270° Longitude

Astronomer Philip James said, "The Hubble results show us that the Viking years are not the rule, and perhaps not typical. Our early assumptions about the Martian climate were wrong."

Imagine this interplanetary weather forecast, brought to you courtesy of the Hubble Space Telescope: "The weather on Mars will be cool and cloudy today. No sign of dust storms. Mostly sunny afternoons with high clouds."

It sounds like a perfect spring afternoon, and indeed it was springtime in Mars's northern hemisphere when Hubble took the pictures that forever changed astronomers' ideas about the weather on this planet and the possibilities for colonization. As it turns out, the 1976 Viking probes visited Mars during a time the planet was having major dust storms. The storms left particles of red dust suspended in the Martian atmosphere, which is why the planet looked so reddish. When the particles were warmed by the sun, the temperature on Mars rose, though it remained well below freezing. But Hubble's pictures show that the dust storms are seasonal and, in part, due to Mars's elliptical orbit. Dust storms prevail when Mars is closest to the Sun, and the atmosphere cools dramatically when its orbit takes it farthest away from the Sun's warming rays. The dust then freezes and settles to the ground as Mars's temperature drops dramatically. Now that the dust has settled down, Mars is becoming even cooler. HST also shows that water in the Martian atmosphere has frozen into ice-crystal clouds, allowing the planet to become drier, too. HST astronomer Steven Lee says, "There has been a global drop in temperature. The planet is cooler and the atmosphere is clearer than seen before." Speaking about the Viking probes he added, "We just happened to visit Mars when it was dusty, and now the dust has settled out."

Like Earth, Mars has an ozone layer in its atmosphere that protects the planet from the Sun's radiation. However, though the ozone layer on Mars is building up and expanding, it is still so thin it would not protect future human explorers from the Sun's harmful ultraviolet rays. There is talk in the space community about the possibility of eventually establishing a colony on Mars. If that happens, early

In this image, Venus is bathed in ultraviolet light. It was 70.6 million miles from Earth when the photograph was taken. The clouds you see permanently cover Venus's volcanic surface. They are made of sulfuric acid, rather than the water vapor clouds found on Earth.

colonists will definitely be wearing space suits, for the Martian atmosphere will not support human life.

Weather information is important to the safety of any journey, whether it's to outer space or across town. If humans are ever going to visit Mars, it will be very important to understand how the weather affects this beautiful planet. No one would want to plan a trip in the middle of one of its dust storms! And it is equally important to understand more about Mars's atmosphere if future spacecraft are going to land there safely. Astronauts need to know the atmosphere's temperature and the distance it extends out from the planet's surface in order to determine how much fuel their spacecraft will need to slow down and enter the planet's orbit in preparation for landing. The weather information HST is gathering is not just for fun. It provides important information to scientists planning space travel for the future.

Unfortunately, the interplanetary forecast for Venus is even more unpleasant than the one for Mars. Imagine the following weather report: "The forecast for Venus: hot, overcast, sulfuric acid showers will continue. Air quality is slightly improved as smog levels subside."

Bright and sparkling Venus is dotted with active volcanoes that erupt and spew sulfur dioxide into the planet's atmosphere. Light from the Sun breaks apart this sulfur dioxide, turning it into acid rain far worse than any that falls here on Earth. On a happier note, Hubble has also shown astronomers that the volcanic activity on Venus seems to be slowing down, so the planet's atmosphere is improving as less sulfur dioxide is spewed into it. Nevertheless, with a surface temperature of 870 degrees Fahrenheit—hot enough to melt some metals—Venus is not a friendly place to visit.

Mars and Venus—along with Mercury, Earth, and tiny Pluto—are called terrestrial planets because they are formed of rocks, dirt, and minerals. "Terrestrial" means "Earthlike." By contrast, the giant outer planets in our solar system— Jupiter, Saturn, Uranus, and Neptune—are not Earthlike. They are largely made up of gases. Like the terrestrial planets, the gas giants have come under Hubble's close scrutiny, too.

On December 1, 1994, when Saturn was 904 million miles from Earth, Hubble photographed the ringed planet and sent fascinating images back to Baltimore. The great telescope picked up evidence of a rare storm on the planet's surface. Packing eastward winds of one thousand miles per hour, the arrowhead-shaped storm spread across an area with a diameter equal to that of Earth—or about 7,900 miles! Hubble was also able to photograph Saturn while looking at its rings edge on, creating an illusion that they had almost disappeared. Saturn's rings are only three hundred feet thick, but their diameter is 171,000 miles across! By looking at the rings edge on in this way, Hubble discovered that

Hubble observed a rare storm on Saturn's surface. The storm is the white arrowhead-shaped feature near the planet's equator. Saturn was 904 million miles from Earth when this picture was taken.

Saturn actually has more moons than previously thought. More study is needed to be certain; however, most astronomers now think that Saturn's moon count should be moved up from eighteen to twenty-two, which would raise the total moon count in our solar system from sixty to sixty-four.

While discoveries about the weather on Mars, the volcanoes on Venus, and the new moons of Saturn are interesting and certainly add scientific information to our knowledge of the solar system, for sheer excitement nothing can beat the front-row seat Hubble offered astronomers as fragments of comet Shoemaker-Levy 9, or SL-9 for short, slammed into Jupiter.

December 1994

May 1995

Once every fifteen years Earth passes through Saturn's ring plane. The top picture was taken by Hubble on December 1, 1994. It shows the familiar ring pattern. By May 22, 1995, the planet's ring system had turned edge on. Viewed this way, the rings seem to disappear and two of its moons are visible.

CHAPTER THREE

A COSMIC COLLISION

Jupiter is named after the Roman god who hurled lightning bolts down on Earth when he was unhappy. The approach of comet Shoemaker-Levy 9 seemed to turn the tables on the god's namesake planet. Now Jupiter was going to be hit with a jolt—not from lightning bolts, but by space debris in the form of a comet named after its three discoverers, astronomers Carolyn and Eugene Shoemaker and amateur astronomer David Levy. As the comet headed on its deadly course, people around the world wondered what would happen to Jupiter. Some were worried; others, like Carolyn Shoemaker, who has discovered more comets than anyone else in the world, were excited at the prospect of an up-close look at a cosmic collision.

Astronomers believe comet SL-9 began its life in the early days of our solar system, some 4.6 billion years ago. At that time a collection of space debris—ice, rock, and metal—formed into a loose ball and kept to itself in an orbit somewhere far beyond Pluto. Around sixty-five years ago, the comet's orbit was disturbed and it began a series of loops in space that eventually put it into a deadly orbit around Jupiter. Finally, on July 7, 1992, the comet came too close. Jupiter's gravity pulled SL-9 apart, and twenty-three chunks of the comet headed for a direct hit on the planet. "If you're a dinosaur on Jupiter, it's time to get

This picture of Jupiter was taken on October 5, 1995, when the planet was 534 million miles from Earth. The arrow points to the site where the unmanned Galileo probe was scheduled to enter Jupiter's atmosphere. At that point 250 MPH winds are sweeping clouds across the predicted entry site.

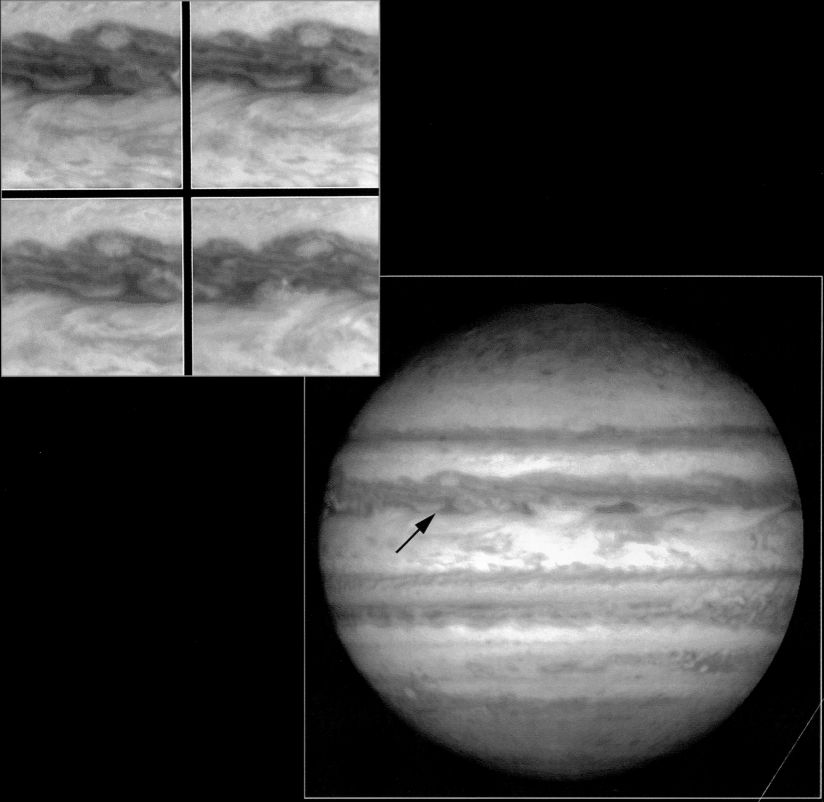

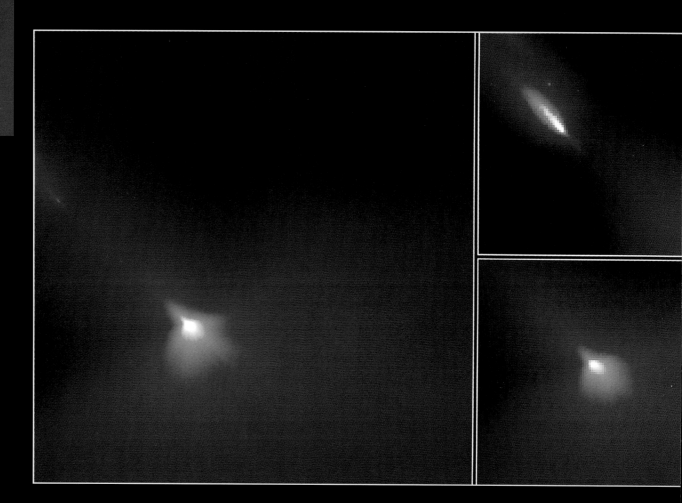

out," said NASA scientist Edward Weiler, as he and other excited astronomers anticipated the event. Dr. Weiler was referring to the fact that many scientists think a similar, but much smaller, comet impact destroyed Earth's population of dinosaurs sixty-five million years ago. Astronomers said one fragment of SL-9 hitting Earth would have unleashed the energy equivalent of a thousand of the world's largest hydrogen bombs. Had that happened, Carolyn Shoemaker said, "It would be the end of civilization as we know it." As they anticipated the collision, everyone wondered about the effects on Jupiter, but everyone was happy Earth was not the landing zone!

During the week of July 18, 1994, scientists at the Space Telescope Science Institute once again crowded around computer consoles, eager to watch the historic event. NASA broadcast pictures over national television. Reporters from around the world wrote stories in newspapers and magazines. This collision between two members of our solar system was destined to be one of the most spectacular astronomical events ever witnessed by humans. One by one during that week, twenty separate fragments slammed into the planet. The space telescope's Wide Field Planetary Camera II, or WF/PC II (pronounced "wiff-pic two" by astronomers), snapped pictures of the impact, as black

debris was flung high into Jupiter's atmosphere, then settled down on top of its clouds. Another instrument aboard the telescope, the Faint Object Spectrograph, or FOS for short, was also at work during the collision. The FOS does not take beautiful pictures; instead, it sends its information to astronomers in the form of a graph. The FOS was able to detect that the black debris from Jupiter contained large amounts of sulfur and ammonia, and it also picked up evidence of silicon, magnesium, and iron, which came from the comet pieces themselves. Hubble also showed us that Jupiter's high altitude winds blow from its poles toward its equator, an intriguing fact since it is just the opposite of how winds blow on Earth. Hubble provided so much data that after the first four days of the impact, Eugene Shoemaker said, "I feel as though I am drowning in a fire hose of information."

There was some debate among astronomers whether SL-9 was a comet or an asteroid. Comets contain water and are usually distinguished from asteroids by their long wispy tails of gas and dust.

Excited astronomers, members of the Hubble Space Telescope Comet Impact Team, gathered around computers at the Space Telescope Science Institute as they awaited the first impacts on Jupiter.

Since SL-9 shared many characteristics of both, astronomers met after the impact to debate what it was—comet or asteroid? After studying the Hubble information and discussing it thoroughly, most astronomers agreed that Shoemaker-Levy 9 was indeed a comet.

Although Jupiter does not appear to have sustained any permanent injury from the collisions, Hubble is still watching the planet and scientists are still studying the information the telescope provided. The great comet collision of 1994 was an extraordinary demonstration of the dramatic and deadly forces that exist in our solar system and beyond.

Of course comets are not the only dramatic objects whizzing through space. There are asteroids and meteoroids, too. Like comets, asteroids and meteoroids are forms of space debris left over from the formation of our solar system about 4.6 billion years ago. Asteroids are large chunks of rocky debris. At least a million asteroids, each over a mile in diameter, orbit the sun in a wide belt between Mars and Jupiter. Smaller chunks of space debris—often no larger than a grain of sand—are called meteoroids. When gravity pulls meteoroids into Earth's atmosphere, they burn across the night sky, becoming meteors. We often refer to meteors as shooting stars, though they are not stars at all but streaks of light in the night sky caused by the heat and friction of Earth's atmosphere upon the meteoroid itself. Occasionally, a meteoroid will survive its fiery ride through the atmosphere and come crashing to Earth as a meteorite. Most meteorites land in the world's oceans, but about 150 smash into dry land each year. However, very few are ever found and examined.

Science has known about asteroids, meteoroids, and meteorites for about two hundred years. In 1801, an Italian astronomer and mathematician, Giuseppe Piazzi (1746-1826), discovered the first and largest asteroid and named it Ceres. Ceres is almost as big as the state of Texas, measuring 604 miles across! Between 1802 and 1807, three more asteroids were discovered—Pallas, Juno, and Vesta. Vesta, the brightest asteroid in the sky and the only one visible with the naked eye, is 325 miles in diameter. Astronomers are particularly interested in Vesta, because part of it landed on Earth as a meteorite and is available for scientific study.

With the exception of meteorites and moon rocks returned to Earth by the astronauts of the Apollo program, astronomers have had no opportunities to study actual pieces of celestial material. And even though many meteorites have crashed to Earth, astronomers have not been able to document where most came from. This is not the case with Vesta. In October 1960, a part of Vesta—in the form of a meteorite—arrived on Earth. Two fence workers from Australia observed the fireball in the sky as it plunged toward the ground. Fortunately, they were not hurt! The Australian meteorite was recovered, and scientists continue to study it. Now Hubble has taken pictures of the celestial portion of Vesta as it orbits the Sun. As a result, Ben Zellner, an astronomer using the Hubble Space Telescope to study Vesta, thinks the asteroid deserves a promotion in classification. He says, "The Hubble observations show that Vesta is far more interesting than simply

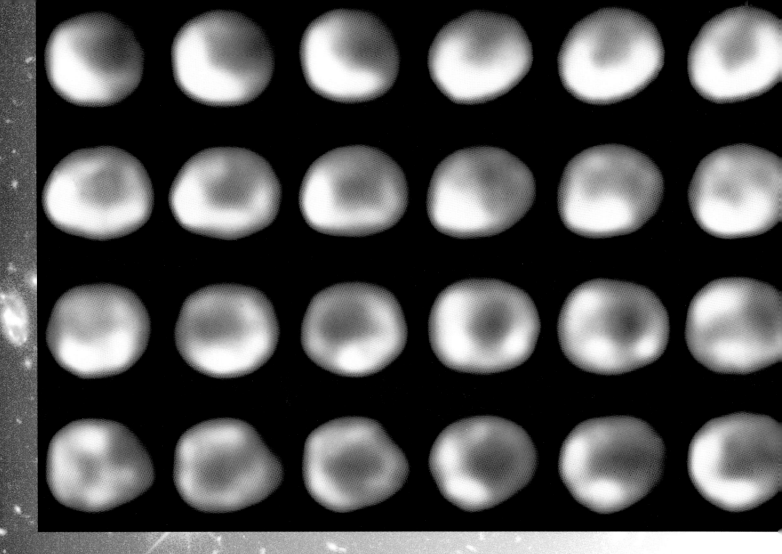

The asteroid Vesta, as seen over a period of 5.34 hours with the Hubble Space Telescope. Vesta has a 325-mile diameter, and although these images were taken when it was 156 million miles from Earth, scientists can detect features as small as fifty miles across.

a chunk of rock in space, as most asteroids are. This qualifies Vesta as the 'sixth' terrestrial planet."

No larger than the state of Arizona, Vesta offers clues to scientists about the origins of our solar system and the interiors of rocky planets like Earth and Mars. Ben Zellner says, "Vesta has survived essentially intact since the formation of the planets. It provides a record of the long and complex evolution of our solar system." The pictures Hubble has taken of Vesta show evidence of lava flows from ancient volcanoes and a giant impact basin where Vesta endured a massive hit from another asteroid. The basin is so deep that Vesta's interior, or mantle, is exposed. A mantle is the part of a planet that lies between its crust and its core. Fortunately for us, no meteorite in human history has ever impacted Earth and left a crater so deep it exposes Earth's mantle. Our understanding of what lies below Earth's crust comes from theory and inferences, not direct observation. Thanks to the Hubble, Vesta's massive crater has been seen for the first time, giving astronomers opportunities to study the once molten interior structure thought to be very similar to Earth's.

Hubble's fascinating pictures and graphs are allowing all of us to gain a better understanding of how things were, and how things will be, in our portion of the universe and beyond.

A meteori
the crust o
fragment v
pounds an
Australia.

5 cm

CHAPTER FOUR

A STAR IS BORN

At the moment the universe was born, scientists believe, the essential ingredients for everything that is—or ever will be—were compressed into one tiny particle that goes by several names, including "the cosmic egg" and "singularity." According to this theory, that single particle existed first. Then, at the beginning of time as we understand it, the particle exploded. Scientists call this explosion the big bang, and with it the universe began to be created. Within one hundred seconds after the big bang, the two lightest and most abundant elements in creation—hydrogen and helium—were formed. Scientists believe that the elements that make up planets, such as iron, calcium, magnesium, and oxygen, followed later, but all of the raw materials expanded outward from that initial explosion.

To picture this kind of expansion, it might help to imagine a new balloon with dots drawn close together on its surface. The dots represent the elements that were created soon after the big bang. As you blow up the balloon and it expands, the dots spread out across its surface, becoming farther and farther apart as the balloon gets bigger and bigger. Scientists, beginning with Edwin Hubble, believe the universe expanded in a much more complicated, but nonetheless similar, way.

Eventually, the atoms of hydrogen and helium were pulled

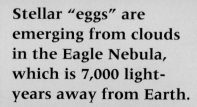
Stellar "eggs" are emerging from clouds in the Eagle Nebula, which is 7,000 light-years away from Earth.

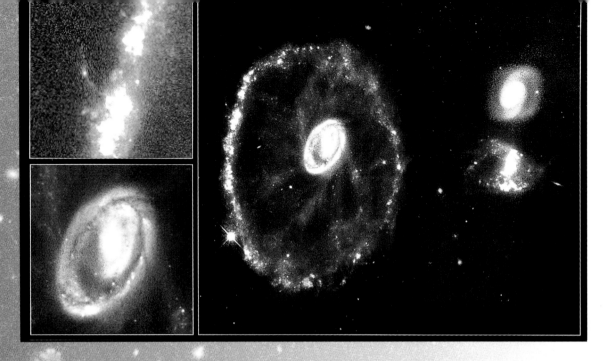

together by gravity to form spinning clouds of space gas called nebulae. These nebulae, or space clouds, were (and are) loosely scattered
throughout the universe. Space clouds are not thick, like the ones that
dot Earth's atmosphere. Instead, the atoms that form a nebula are
widely spread apart, making these clouds extraordinarily thin and
light—much lighter than the air you breathe. To picture how thin the
space gas is, consider that a cup of the air you breathe contains a
quadrillion (a thousand million million) atoms. A cupful of space gas
contains only about ten thousand atoms. Amateur observers just using
binoculars can sometimes see one of these glowing nebulae along the
track of the Milky Way. The Hubble Space Telescope, however, provides a much better view!

Stars begin their life deep inside a nebula. Their birth starts with
a struggle between two forces—heat and gravity. If heat wins the
battle, the spinning nebula will expand, becoming a bigger and bigger
space cloud. No stars will be born in it. On the other hand, if gravity
wins the battle, the nebula will be pulled in on itself, spinning faster
and faster as the pressure increases. Eventually, it will break up into

Opposite page: A ring of stars is born as a result of the collision of two galaxies. The Cartwheel Galaxy is located 500 million light-years away from Earth. Scientists do not know which of the two smaller galaxies to the right of ring was the intruder.

thousands of fragments. A star is formed when a fragment of that space gas contracts, forming a ball of gas called a protostar. As gravity continues its pressure, the new ball of gas gets hotter and hotter. Finally, when the temperature inside the newly forming star gets to be around 18 *million* degrees Fahrenheit, hydrogen inside the ball is fused into helium. When that fusion happens, energy is released in the form of heat and light, and a star is born. The entire process of star formation process takes millions of years to accomplish.

Stars are often born in clusters at about the same time and place. Sometimes clusters of stars are created when galaxies collide. If that happens, the shock waves from the intergalactic collision send ripples of energy into the surrounding nebulae. The shock waves may compress the gas and dust in the nebulae, and the process of star formation starts all over again.

The Hubble Space Telescope has provided astronomers with astonishing pictures of this process of star birth in the Cartwheel Galaxy. The bright ring that circles the galaxy contains blue knots that are gigantic clusters of newborn stars. The diameter of this ring of new stars is 150,000 light-years across; our entire Milky Way would fit inside it! Before the collision, astronomers believe, the Cartwheel Galaxy was spiral-shaped, like the Milky Way, but the collision temporarily disrupted the galaxy's shape. Thanks to Hubble's capabilities, astronomers can see the reemergence of spiral arms forming around the galaxy's nucleus. Although most galaxies are spiral-shaped, others are elliptical, or egg-shaped. Still others have no distinct shape at all and are called irregular galaxies.

For close-ups of star formation, nothing is more spectacular than the images Hubble returned to Earth of the Eagle Nebula, seven thousand light-years away from Earth, but still in the Milky Way. Dr. Jeff

Hester used the Hubble to study that nebula. He said, "We are uncovering the environment of star formation and getting it out in the open, where we can see so much of what is going on." Hubble photographed dense pillars of gas protruding like stalagmites from a dark cloud of hydrogen gas. If you look carefully, you can see small blisters or bumps on the ends of fingerlike protrusions from the surface of the cloud. Each one of these "fingertips" is larger than our entire solar system! The bumps are called evaporating gaseous globules, or EGGs, and stars are forming inside them. The EGGs have been exposed by a process called photoevaporation. Photoevaporation occurs when intense ultraviolet light coming

from nearby hot, young stars evaporates the gas surrounding the EGGs.

Dr. Hester explained that process by saying, "It's a bit like a wind storm in the desert. As the wind blows away the lighter sand, heavier rocks buried in the sand are uncovered. The ultraviolet light from nearby stars does the digging for us, and we study what is unearthed. In some ways it seems more like archaeology than astronomy." It is the first time anyone on Earth has ever witnessed this process. Because the gas that nourishes new stars has been evaporated, some of these EGGs may never produce stars.

Hubble has confirmed that a dusty disk forms around a newborn star. New pictures show jets of gas

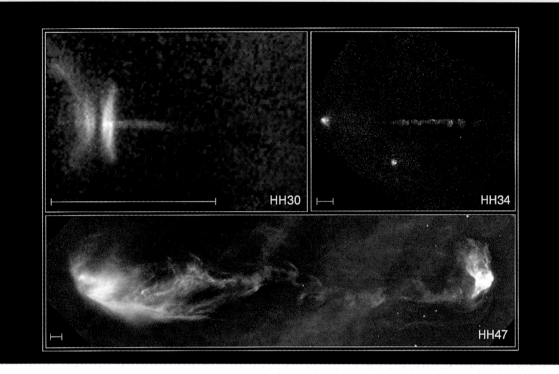

HH30

HH34

HH47

The jet of gas ejected from this newly forming star is three million miles long!

blazing away from the surface of these new stars. Although these jets blaze for a relatively short period of the star's life—one hundred thousand years or so—they are crucial in determining the new star's life span. The jets appear to control how much space dust will fall onto the star. The more dust, the more massive the star. Size is significant in star birth, because a star's lifetime is determined by its mass. Big, massive stars tend to build up quickly, but they die quickly, too. However, even the shortest-lived stars have a life expectancy of several million years!

Although astronomers know that new stars are being born throughout the universe every day, there is still much to learn about when and how it happens. The Hubble Space Telescope is helping astronomers understand the process of star birth and growth throughout the vastness of the universe, where billions of stars, in billions of galaxies, are being created all the time.

Opposite page: Pillars of cool hydrogen gas protrude like stalagmites from the "floor" of an interstellar cloud. The tallest pillar (left) is about one light-year long, or about 5.9 trillion miles from base to tip. Stars are embedded in the EGGs at the tip of these pillars.

CHAPTER FIVE

A STAR DIES

Jeff Hayes says, "Our Sun is an ordinary star, but if it blinked out tomorrow, every living thing on Earth would have 8.3 minutes left to say good-bye." It takes the Sun's light 8.3 minutes to reach Earth, and although life on Earth would stop without that sunlight, no one expects the Sun to die quickly (or soon) so there is no need to worry. Our Sun is still a vibrant, middle-aged star, destined to supply our planet with light and heat for another 5 billion years or so.

However, like all stars, the Sun will eventually die. And—similar to the death of other stable stars—the process will take place over a long period of time. Right now, the Sun is converting 4.5 million tons of mass into energy every second. That is enough energy to supply the electrical needs of a country like the United States for the next 50 million years! This kind of heat and energy at the center of our Sun (and other stars like it) pushes its gases outward. Gravity, however, pulls the gases back toward the star's center, so that despite the push and pull, the Sun has a stable size. After a turbulent birth, most stars, like the Sun, settle into eons of middle-aged life, with their gases carefully balanced between expansion and contraction.

Toward the end of an ordinary star's life, its hydrogen supply begins to be used up. The hydrogen burns faster and faster, producing

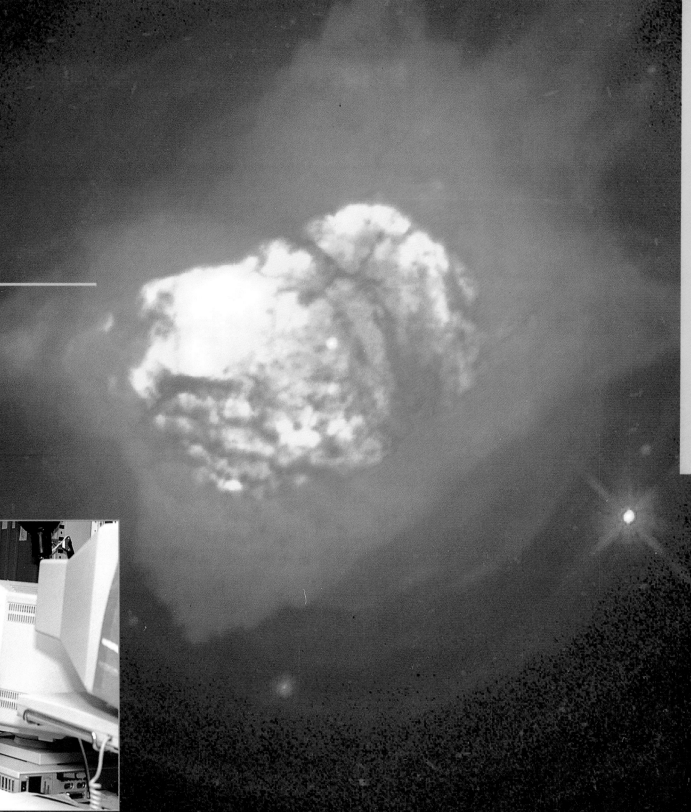

Opposite page: Jeffrey Hayes uses a computer at his office at the Space Telescope Science Institute in Baltimore, MD. Part of Jeff's job is to help astronomers plan observations using the Hubble Space Telescope.

This page: An extraordinary Hubble photograph of the death of a star similar to our Sun. The photo shows the outer layers of material being ejected into space. The small dot in the center of the image is the core, or white dwarf star.

Size of Star

Size of Earth's Orbit

Size of Jupiter's Orbit

extra energy which causes the dying star to swell. Some day, our Sun will swell to fifty times its original diameter, or a thousand times bigger than it is now. Daytime temperatures on Earth will reach 2,600 degrees Fahrenheit, but life on Earth will have stopped long before this happens! As the bloated Sun begins to cool down, its color will change from yellow to red, and it will be known as a red giant. Red giant stars do not have enough gravity to hold onto their outer layers. These layers are blasted into space and become planetary nebulae. Eventually, the star's central core of helium shrinks into a kind of star known as a white dwarf. Although astronomers have understood how this process works for years, the Hubble Space Telescope has provided them with amazing photographs that show the process in detail.

Ordinary stars like the Sun are born slowly, live a long time, then die slowly in the process just described. But other kinds of stars go out with a bang. The word *nova* comes from the Latin word that means "new," but it actually describes something that can occur near the end of a star's life. A nova forms when two stars—one a middle-aged red giant and the other an old white dwarf—orbit each other, creating a binary (or double) star system. Since the white dwarf has far more gravity than the red giant, it pulls the red giant's hydrogen toward itself. This process makes the white dwarf get bigger and bigger, building up more and more mass. At this point, the star may get rid of its excess matter with tremendous explosions that propel light, heat, and elements out into space. An exploding nova can shine as brightly as one hundred thousand Suns, but it fades quickly, too. On the other hand, the white dwarf could continue to grow and even reach a size known as Chandrasekhar's limit, which is 1.44 times the mass of the Sun. At this limit, the entire star collapses and explodes as a supernova. The energy from a supernova can equal the energy from an entire galaxy—a billion or more stars.

Eta Carinae is one of the largest and most massive stars ever seen, and Hubble's picture of it is a favorite of Laura Danly's. She said, "Eta Carinae is my favorite image from Hubble. The telescope lets me feel as if I have traveled close to it. Of course, the entire universe is beautiful, and Hubble lets us see the glory of it all." Located ten thousand light-years from Earth, Eta Carinae is 150 times bigger than our Sun and four million times brighter. However, unlike the Sun, Eta Carinae is not stable. It is prone to violent explosions and astronomers think it is on the brink of destruction.

In another spectacular image, HST photographed hula hoop–like double rings of glowing gas that surround a star with the unimaginative name of Supernova 1987A. Supernova 1987A is located 169,000 light-years away, in another galaxy called the Large Magellanic Cloud.

Opposite page: This is the first direct image of Betelgeuse, an enormous red giant star in the shoulder of the constellation Orion the Hunter. If Betelgeuse replaced the Sun at the center of our solar system, its outer atmosphere would extend past the orbit of Jupiter! This image is the equivalent of seeing a car's headlights from 6,000 miles away.

Dr. Christopher Burrows, an astronomer with the European Space Agency and the Space Telescope Science Institute, used the great telescope to study Supernova 1987A. No one was more astonished than he when the rings around this supernova were revealed. Nothing like them had ever been seen in space before! No one knows whether these hoops of gas existed before the star exploded or whether they were created at the time of the star's death. Astronomers who have studied the images have not agreed among themselves. "The Hubble images of the rings are quite spectacular and unexpected. This is an unprece-dented and bizarre object. We have never seen anything behave like this before," said Dr. Burrows, who is still studying the informa-tion Hubble has provided.

On Earth living things die, decompose, and enrich the planet, and new life can begin again. The same process is repeated throughout the universe as stars die and eject material that will be used to create future gener-ations of stars, planets, and perhaps life itself. Laura Danly studies the nebulae in which stars are born and die, and she says, "Every atom has its own story. Clouds of space gas determined the evolution of the Sun, which

Opposite page: Astronomers believe Eta Carinae, a star 150 times larger than our Sun and more than 4 million times brighter, is on the brink of destruction. Eta Carinae is 10,000 light-years away.

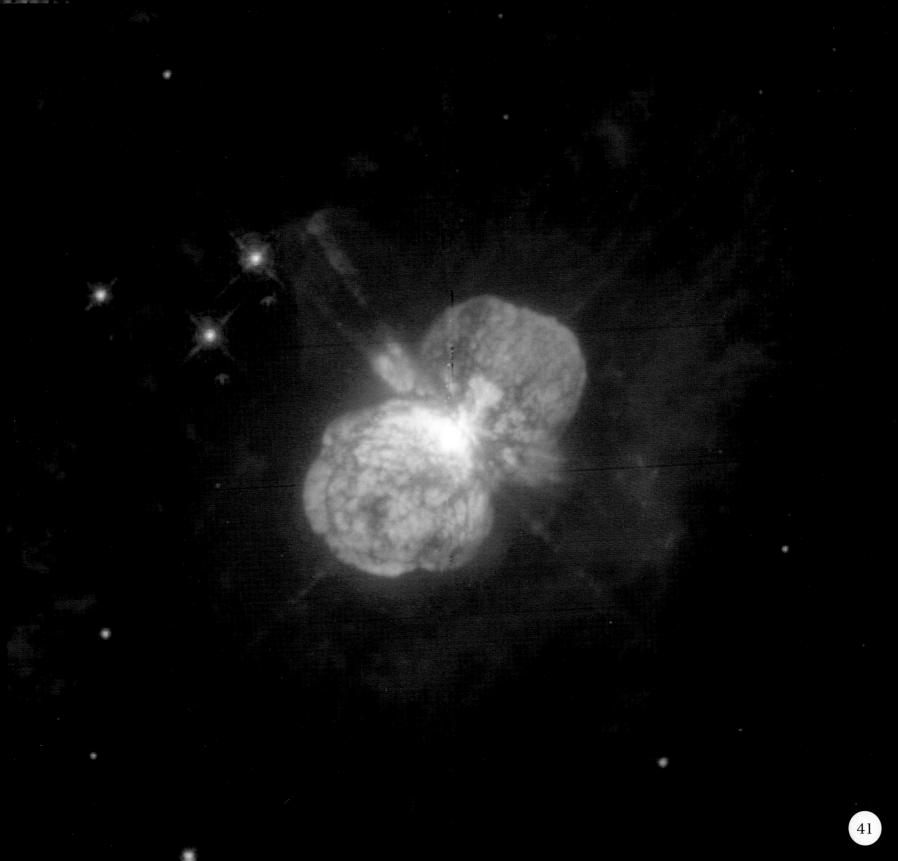

HST captured an image of mysterious rings surrounding the supernova 1987A, 169,000 light-years away. Dr. Christopher Burrows was surprised by this sight, since based on previous observations his team expected to see an hourglass-shaped bubble of gas thrown into space by the dying star.

In this image, the central star was a red giant only a few hundred years ago. It is now surrounded by a cocoon of dust, which is darker in the center and hides the dying star from our view.

determined the evolution of Earth. The atoms in each nebula will determine the evolution of other stars in the future." The atoms that make up the material from which our entire solar system was formed were once inside a star. The Earth and everything in it is literally made from star dust.

CHAPTER SIX

BENDING SPACE AND TIME

When a sock, homework, scissors—or anything else—disappears, never to be seen again, people often say, "It's fallen into a black hole." The term has become part of our everyday vocabulary, though it has only existed since 1967. That was the year American astronomer and physicist John Wheeler coined the term black hole to describe an astronomical mystery that had puzzled scientists for about two hundred years.

In order to begin to understand these mysterious objects, you must understand something about Albert Einstein's famous general theory of relativity. Very simply put, that theory says that gravity affects the actual shape of space and time—it bends or warps it. To understand how a gravitational field is bent, think of the flexible surface of a trampoline. Let it represent space-time. Now put a tennis ball on the trampoline. It does not make a measurable difference on the surface. But if *you* stood on the trampoline, its surface would bend inward, creating a depression or hole around your feet. The trampoline's surface bends noticeably because you have more mass and therefore more gravity than a tennis ball. In fact, if the tennis ball were close enough to you, your gravitational pull would force it to roll toward your feet. Although this example lets you see how gravity

An artist's rendition of a supermassive black hole.

(you) bends space (the surface of the trampoline), the depression your body makes in the trampoline is a gravitational hole, not a black hole. To be a true black hole, the depression in the trampoline would have to be so deep that neither the tennis ball nor you could ever be retrieved from it, and even light would be sucked in.

A black hole is not a hole exactly. It is a region of space that is remark-  able because it is so dense. Scientists believe that black holes are created when extremely massive stars—at least eight times larger than our Sun—die. As these stars die, they begin to collapse in on themselves. They send their outer layers into space in the form of supernova explosions, but their cores remain. If the core has more than three times the mass of the Sun, nothing in the universe will be able to stop it from collapsing indefinitely. The star becomes incredibly small and dense. And its gravity becomes so strong and so concentrated that nothing inside a boundary known as the *event horizon* will ever escape.

It is difficult to escape the pull of gravity on Earth, but it can be done if the object making the escape is traveling fast enough in the right direction. For example, when NASA wanted to put the Hubble Space Telescope into Earth orbit, they placed it inside the shuttle *Endeavour* and programmed the space plane to reach a speed of 18,000 miles per hour (that's 5 miles a second!) after launch. Earth's gravity pulled *Endeavour* downward, but since the shuttle was going so fast, the surface of Earth curved away from it at the same rate it fell. The shuttle fell toward Earth all the way around the planet, without ever hitting it;

that's what it means to be in orbit.

However, when NASA sent men to the Moon, their Apollo spacecraft had to go even farther from Earth's surface. That took more speed. Our moon-walking astronauts had to reach a speed of 25,000 miles per hour before they could leave Earth behind. Scientists call that 25,000 mile-per-hour speed Earth's escape velocity. All planets have an escape velocity—a speed at which an object will leave the planet altogether rather than fall back to its surface. The more massive a planet, the higher its escape velocity. For example, the Moon has only one sixth as much gravity as Earth. When the astronauts left the Moon and returned to Earth, their spacecraft only had to reach a speed of 5,300 miles per hour before it escaped the Moon's gravitational hold. Five thousand three hundred miles per hour is the Moon's escape velocity.

The more massive an object is, the faster you have to go to escape its hold. Nothing in the universe travels faster than light, which moves at the breathtaking speed of 186,000 miles per second. Light has little trouble escaping from the gravitational hold of stars like our Sun, but even light itself cannot escape from a black hole.

Nevertheless, if you were traveling through space and you came close to a black hole, you would not necessarily be pulled into it. Although black holes emit tremendous gravitational pull, the farther away you are from the center of that pull, the less force it exerts. You would have to cross the black hole's boundary, the event horizon, before you were sucked inside for good. The event horizon is an imaginary spherical

Opposite page: Albert Einstein was considered a bad student but he went on to develop the theory of relativity and was hailed as a genius.

surface that marks the edge of the black hole. Along that edge, the escape velocity equals the speed of light. Once inside, in order to get away from its pull, you would have to be traveling faster than the speed of light. And that's impossible. Therefore, nothing that crosses the event horizon ever escapes. Outside the event horizon it's another story. There the escape velocity—though amazingly high—is still less than the speed of light. Theoretically, at least, if your rockets were strong enough, you could overcome the pull of gravity and you could avoid being sucked into the black hole.

All laws of physics, as we understand them, break down inside a black hole. Anything drawn into a black hole disappears forever, and no one knows with certainty exactly what lies within. In his famous book, *A Brief History of Time: From the Big Bang to Black Holes*, the brilliant British astrophysicist Stephen W. Hawking says, "Anything or anyone who falls through the event horizon will soon reach . . . the end of time."

Since the gravity in a black hole is enough to prevent light from escaping, no one has ever seen an actual black hole. But no one has ever seen the wind, either, yet we know it is there by observing the effect it has on other things. When leaves flutter in the trees or kites fly upward, we know the wind is present. We can know a lot about the wind without ever seeing it. There are scientific instruments that measure its speed and direction, and it is the same with black holes. Scientists look for the presence of black holes by observing the effect they have on the things in space around them.

Because of their tremendous mass and gravity, black holes cause

An artist's rendition of what a cross-section of a massive black hole might look like. Particles and radiation blast away from the hole in twisting spirals of energy that create extragalactic jets thousands of light-years long. The black dot in the center is the hole's event horizon. Nothing that crosses it will ever escape.

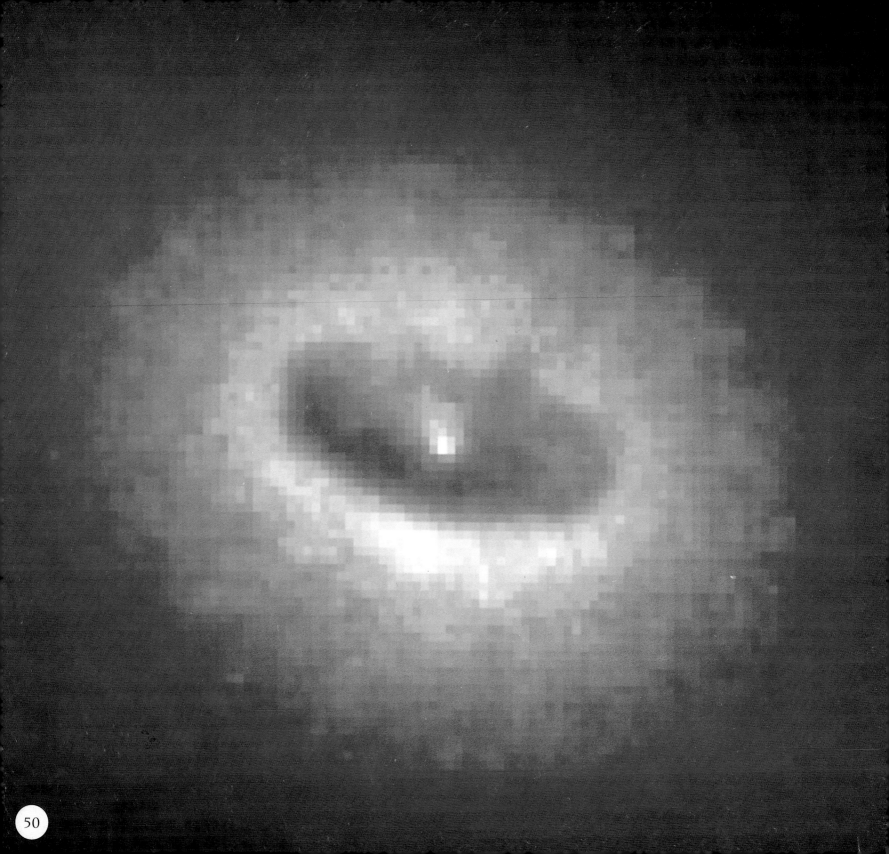

stars and gases to orbit them, just outside the event horizon, at incredible rates of speed. Jeffrey Hayes said, "We can measure all this stuff that's whizzing around. It's really quite amazing."

In order to measure motion in the universe, Jeff and other astronomers who work with the Hubble use the Faint Object Spectrograph, or FOS for short. The FOS can separate starlight into the colors of the rainbow, also known as the spectrum—what Sir Isaac Newton studied with his prisms. Astronomers know that when an object is moving away from us, its light appears redder. When it moves toward us, its light appears bluer. They talk about light being redshifted or blueshifted as they try to measure the speed of the object that emitted it. By using Hubble's FOS, astronomers at Space Telescope Science Institute were able to measure the speed of the gas rotating around a suspected black hole in the center of galaxy M87. It is moving 1.2 million miles per hour, indicating that there is a tremendous gravitational pull in the center of this galaxy. There are not enough stars there to account for this kind of gravity. The evidence of swirling gas points to only one other possibility—a large black hole.

Opposite page: HST photographed a disk that could be fueling a black hole forty-five million light-years away, in the Virgo Cluster. This image was taken before the Hubble telescope was repaired.

Gas Disk in Nucleus of Active Galaxy M87

Hubble Space Telescope
Wide Field Planetary Camera 2

Another Hubble instrument, the Wide Field Planetary Camera II, took these close-up photographs of a disk of gas swirling around the black hole's event horizon. The diagonal streak in the image is a jet of high-speed electrons produced by the black hole's nucleus, or "engine."

Though no one has ever seen the black hole itself, by measuring and photographing activity near the event horizon, the Hubble Space Telescope has provided the first definite proof that black holes really exist.

Astronomers believe black holes occupy the center of many galaxies—including the Milky Way. They are out there, pulling everything that crosses their event horizons into themselves. Fortunately, Earth is too far away from the event horizon of any of the Milky Way's black holes to be pulled inside. Therefore, the only black holes earthlings need worry about are the ones that suck down their homework!

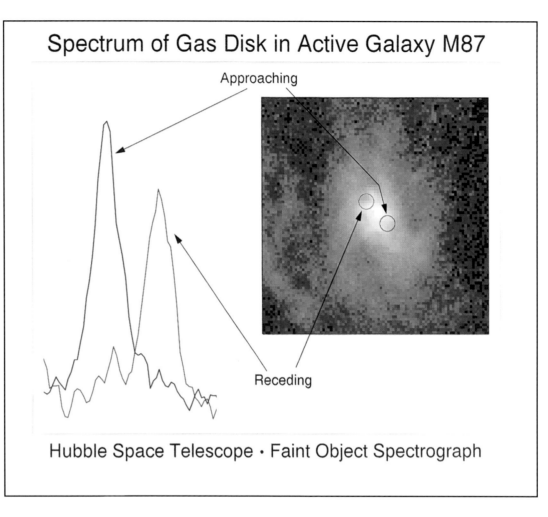

Spectrum of Gas Disk in Active Galaxy M87

Approaching

Receding

Hubble Space Telescope · Faint Object Spectrograph

Scientists use the Faint Object Spectrograph aboard the Hubble to measure the speed of a rotating disk of hot gas in the center of galaxy M87. This image convinced scientists that a black hole lurks at the center of the galaxy.

CHAPTER SEVEN

OTHER SOLAR SYSTEMS? OTHER

Dr. Story Musgrave was in charge of the crew of astronauts who made the repairs to the Hubble Space Telescope. He is a brilliant scientist who holds six university degrees, in addition to being a physician and surgeon. Story Musgrave is a person who thinks before he speaks and does not utter careless comments. As he prepared himself for his mission to repair the telescope, Story admitted to a dream; he wanted to make contact with other intelligent beings in the universe. At that time, in 1993, he said, "It's almost a statistical certainty that there are beings out there millions of years more advanced than we are. The possibilities are absolutely immense." The Hubble Space Telescope has uncovered evidence that increases those possibilities even more.

Thanks to powerful telescopes, astronomers have figured out how stars form in the universe. However, no one had ever observed the birth of a planet. In fact, though astronomers guessed that there might be planets outside our solar system, no one really knew. The scientific community relied on hypotheses to explain how planets were formed. A hypothesis is an explanation of how or why something happens, based on scientific study and reasoning. Scientists begin with a hypothesis, then gather evidence in an attempt to prove whether it's true or not. The process is long and difficult, and often people do not

Opposite page: Four newly discovered protoplanetary disks surround young stars in the Orion Nebula. As they evolve, these disks may go on to form planetary systems like our own. The smallest of the four disks is twice the diameter of our solar system, and the largest is eight times the diameter.

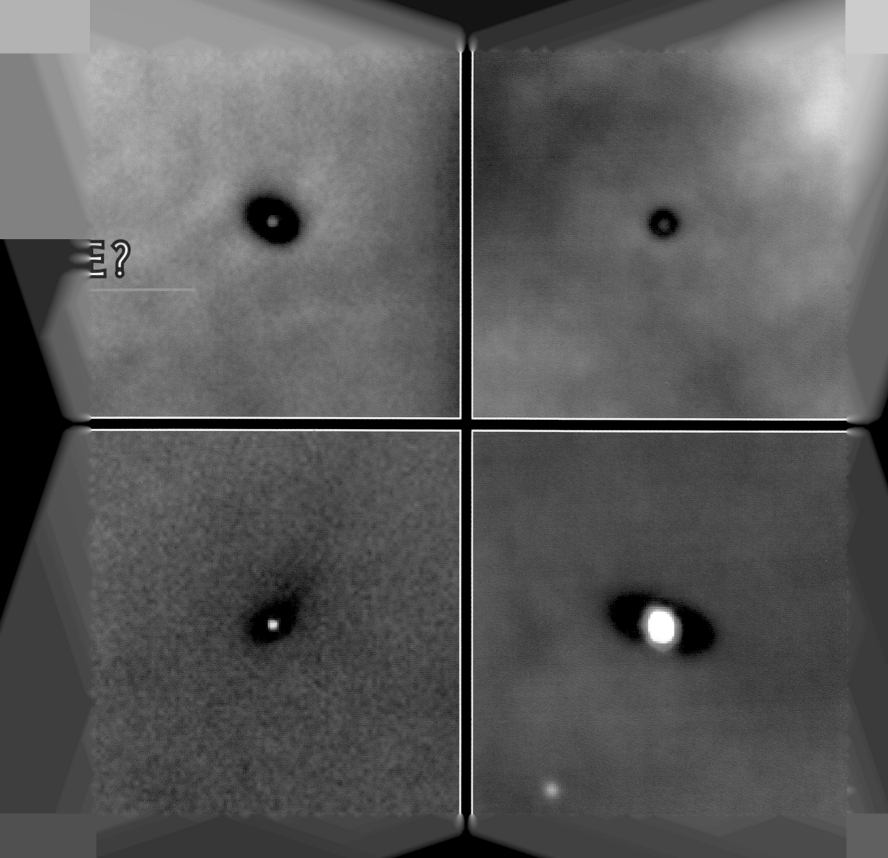

agree about what the evidence indicates. Nevertheless, scientific debate often adds to the total amount of knowledge that is gathered.

For years scientific theory has stated that the planets in our solar system began to form about one million years after our Sun ignited and became a star. As the nebula that gave birth to the Sun started to shrink, it left behind bits of space dust, rock, ice, and gas. Astronomers believe this material orbited the newborn Sun until it eventually formed into a massive, flat ring that they call a protoplanetary disk. The theory goes on to say that this disk was hottest where it was closest to the Sun, and cooler at its outer edges. Gradually, gravity began to pull the pieces of dust, rock, ice, and gas together in clumps. The Sun's heat blasted most of the light gases away from the inner planets, like Earth, leaving them small, rocky, and terrestrial. Farther away—toward the outer edges of the protoplanetary disk—it was cooler. The gas giants of the outer solar system— Jupiter, Saturn, Uranus, and Neptune—kept their thick atmospheres.

Balance played an important role in this theory of planetary formation. Had the collisions between the pieces of dust, rock, ice, and gas been too violent in the early stages, the planets would have shattered as quickly as they were being built. And if our Sun had been too hot, or the protoplanetary disk had been too close to it, the Sun's radiation might have evaporated the disk before anything began to happen. There was much that could go wrong, and an entire solar system orbiting a star seemed that it might be a rare phenomenon. In fact, although astronomers had observed millions of other stars similar to our Sun, no one had ever seen another solar system—or even a protoplanetary disk—until the Hubble was available for astronomers to use.

The Orion Nebula, 1,500 light-years away. Inside this great cloud of space dust and gas, astronomers have finally spotted evidence of planetary formations revolving around other stars.

In 1994, astronomers C. Robert O'Dell and Zheng Wen used HST to study 110 stars in the Orion Nebula, 1,500 light-years from Earth. For the first time in the history of astronomy, they found fifty-six stars that had protoplanetary disks orbiting them! At the time, Dr. O'Dell said, "Since it is easier to detect the star than the disk, it is likely that far more stars are being orbited by protoplanetary material." In fact, by 1995 Dr. O'Dell and his colleagues had found 153 disks circling newborn stars in the Orion Nebula. Scientists theorize that our planetary system formed around the Sun when it was a million years old. Speaking of the stars he is observing with the Hubble and the protoplanetary disks surrounding them, Dr. O'Dell said, "That's just about the age we're seeing here." Thanks to the Hubble Space Telescope and the work of astronomers like Dr. O'Dell and Zheng Wen, the theory of planetary formation now has hard evidence to support it. For now, the observations merely prove the existence of the disk, a kind of planetary incubator. Fully developed planets outside our solar system have not been observed at this time.

So far, Earth is the only place in the universe where life is known to exist. The stars in Orion are very young and their planetary systems are just developing. However, Dr. O'Dell has said the disks he and his colleagues have discovered are likely to contain the same materials that are in the planets of Earth's solar system. That raises the possibility that other planets similar to Earth are out there somewhere—especially when one considers the hundreds of billions of galaxies, each full of hundreds

Opposite page: An edge-on view of a protoplanetary disk in the Orion Nebula. The star is mostly hidden inside the disk. Scientists think our solar system formed from a disk like this 4.5 billion years ago.

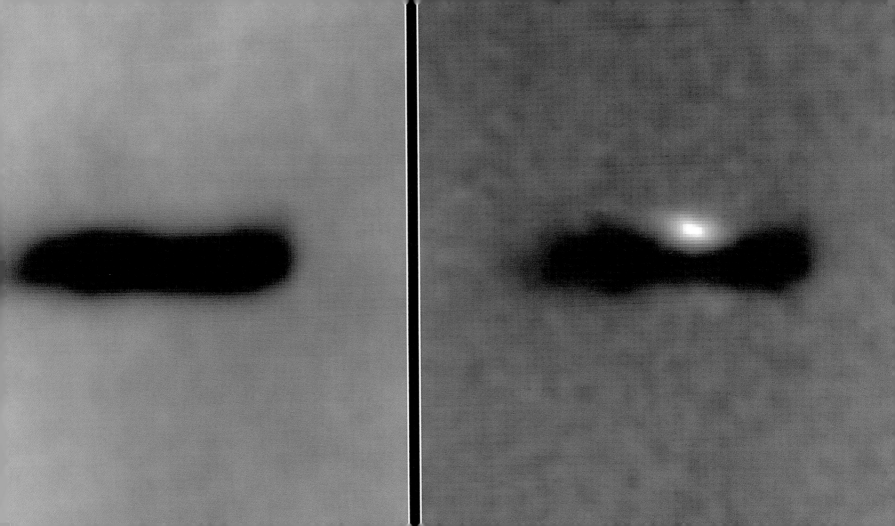

of billions of stars, that are scattered throughout the universe! According to the scientists at Space Telescope Science Institute, the existence of so many young stars with protoplanetary disks mathematically increases the likelihood of other planetary systems being discovered in the universe. Dr. O'Dell says this: "It means the building blocks are there; however, it doesn't mean that planets will certainly form."

The kind of intelligent beings Story Musgrave talked about encountering would have to come from a planet similar to Earth. Like Earth, it would have to be a planet that could store the chemicals that are needed to produce biological life. It would have to have an atmosphere that contained the gases used by life forms, and it would have to have enough gravity to hold its atmosphere to the planet's surface, protecting the life that was growing there. And any life-bearing planet would have to have just enough heat and light from its nearby star. Too much or too little would kill the potential for life—at least, life as we understand it!

No one knows if the universe contains another planet answering this description or not. However, most scientists seem to believe that statistically, at least, it is possible. Whether we are alone in the universe or not, the science done with the Hubble Space Telescope helps us understand that life on our beautiful planet is a rare, delicately balanced gift that we should treasure.

The universe into which all of us are born is an amazing creation, one that is still in the process of evolving. When Laura Danly said, "Every atom has its own story," she was talking about cause and effect. We are who and what we are because certain atoms combined in specific ways to create each unique individual person. Dogs and cats are what *they* are because other atoms combined in other ways to cre-

Opposite page: Five young stars in the Orion Nebula. Four of them are surrounded by protoplanetary disks.

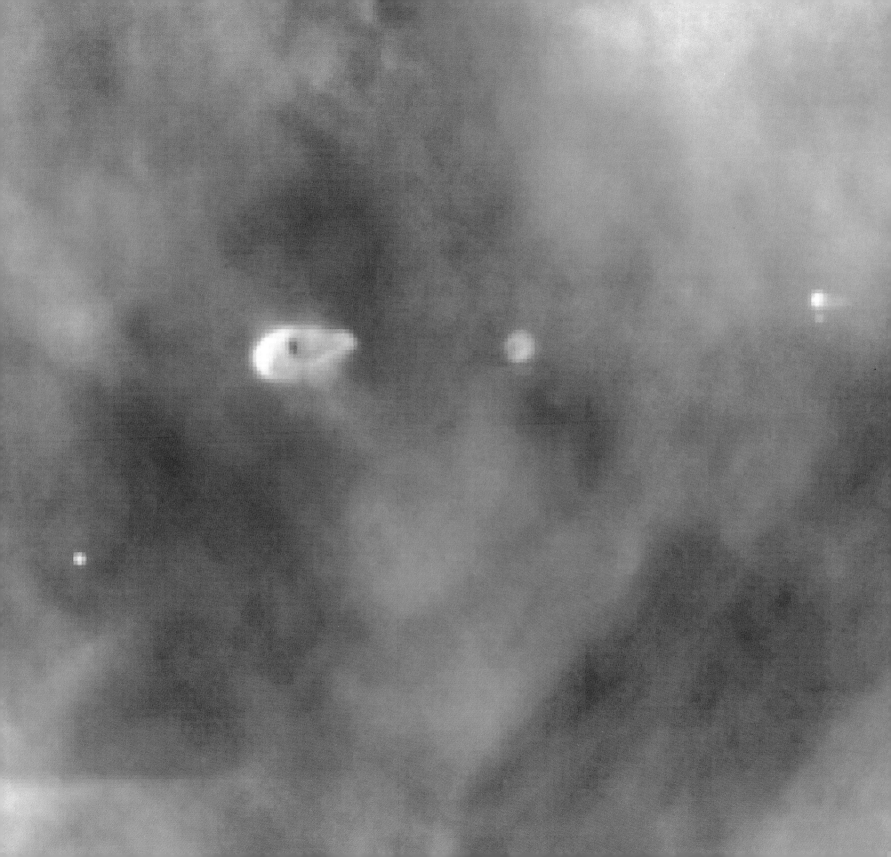

ate them. The stars and the planets and everything else that exists are what *they* are as a result of the creative force at work in the universe. This creation process is still going, and since the beginning of human life we have tried to understand its mysteries. Laura Danly says, "The universe is accessible to us. It's just there, waiting to be understood." And the universe *can* be understood. We can study its raw material and try to understand the laws it operates by. Those laws do not change, though people throughout the centuries have often misunderstood them. In *A Brief History of Time*, Stephen Hawking asks a question: Why does the universe bother to exist? The Hubble Space Telescope is helping scientists explain how things turned out as they did. The world's great religions explain why.

Opposite page: A "deepest-ever" view of the universe. Bluer objects contain young stars and/or are relatively close, while redder objects contain older stellar populations and/or are farther away.

An artist's rendition of a protoplanetary disk of gas and dust circling a newborn star.